SCHOLASTIC

GRADE
3

D0580479

Success With
Multiplication & Division

New York • Toronto • London • Auckland • Sydney
Mexico City • New Delhi • Hong Kong • Buenos Aires

Teaching *Resources*

State Standards Correlations

To find out how this book helps you meet your state's standards, log on to **www.scholastic.com/ssw**

Written by Libby Beck and Terry Winterman
Cover design by Ka-Yeon Kim-Li
Interior illustrations by Carol Tiernon
Interior design by Quack & Company

ISBN 978-0-545-20087-5

Introduction

Parents and teachers alike will find this book to be a valuable learning tool. Students will enjoy completing a wide variety of activities as they practice multiplication and division skills. The activities, which include using codes, solving puzzles, drawing pictures, and much more, are both engaging and educational. Take a look at the Table of Contents and you will feel rewarded providing such a valuable resource for your students. Remember to praise them for their efforts and successes!

Table of Contents

Hockey Talk (Understanding multiplication) 4

Hopping Along (Multiplying by 2s and 3s using a number line) 5

Picture Perfect (Multiplying by 4s and 5s using arrays) 6

Shining Brightly (Multiplying by 4s and 5s) 7

A Wheel of Facts (Review—multiplying by 2–5s) 8

Product Drop (Multiplying by 5–7s) 9

Sweet Success (Multiplying by 6s and 7s) 10

The Product Trail (Multiplying by 8s and 9s) 11

Climbing to the Top (Multiplying by 8s and 9s) 12

The Number Man (Review—multiplying by 2–9s) 13

The Case of the Missing Factors (Review— multiplying by 2–9s) 14

Triple Hit (Multiplying 3 factors) 15

One Step, Two Step (Multiplying without regrouping—2-digit top factors) 16

Times Race (Multiplying without regrouping—2-digit top factors) 17

Carry Carefully (Multiplying with regrouping— 2-digit top factors) 18

Falling for Multiplication (Multiplying with regrouping—2-digit top factors) 19

Let's Review! (Review—multiplying with 2-digit top factors) 20

Snack Time! (Multiplying money) 21

Sweet and Juicy (Multiplying without regrouping—3-digit top factors) 22

Zany and Brainy (Multiplying without regrouping—3-digit top factors) 23

Check It Out (Multiplying with regrouping— 3-digit top factors) 24

Number Fun With Barky (Solving word problems using multiplication) 25

Tic-Tac-Toe (Review—multiplication) 26

Working With Area (Understanding area) 27

What Is Division? (Understanding division) 28

Filled With Stars (Dividing by 2s and 3s) 29

Alien Adventure (Dividing by 2s and 3s) 30

Flying Back (Dividing by 4s and 5s using a number line) 31

Dividing Is a Breeze (Dividing by 4s and 5s) 32

Dividing Evenly (Review—dividing by 2–5s) 33

Leaping Lily Pads (Dividing by 6s and 7s) 34

Field Trip Fun (Dividing by 6s and 7s) 35

The Skate Divide (Dividing by 8s and 9s) 36

Dividing Race (Dividing by 8s and 9s) 37

Fishy Fact Families (Identifying fact families) 38

Over the Hurdles (Dividing with remainders) 39

More Remainders (Dividing with remainders) 40

Figure It Out (Interpreting quotients and remainders) 41

Keep on Dividing (Dividing without remainders— 2-digit dividends) 42

Dividing the Loot (Dividing without remainders— 2-digit dividends) 43

It's All Relative (Finding missing dividends) 44

Division Treats (Solving word problems using division) 45

Decision Time (Solving word problems using multiplication and division) 46

Answer Key 47–48

Hockey Talk

 Multiplication *is repeated addition using equal groups. The numbers being multiplied together are called* **factors**. *The answer is called the* **product**.

Charlie's coach bought 3 bags of hockey pucks. There are 4 pucks in each bag. How many pucks did he buy altogether?

This problem can be solved by drawing a model.

 = 12 pucks

3   x   4   =   12 pucks
factor   factor   product
There are 3 equal groups of 4 pucks.

A. Charlie's team played 3 games. They scored 3 points in each game. How many points did they score in all? Draw a model to show your answer. (Draw a box for each point.)

B. Charlie and his friend each have 4 hockey sticks. How many sticks do they have? Draw a model to show your answer.

 On another sheet of paper, write your own word problem that involves repeated addition using equal groups. Ask a friend to solve it by drawing a model.

Hopping Along

 *A **number line** can be used to help you multiply. One factor tells you how long each jump should be. This is like skip-counting. The other factor tells you how many jumps to take.*

$2 \times 6 = 12$

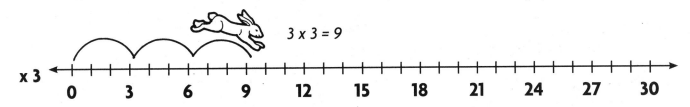

$3 \times 3 = 9$

Use the number lines above to help you multiply by 2s and 3s.

A. $2 \times 2 = $ _____ $3 \times 3 = $ _____ $6 \times 2 = $ _____

B. $4 \times 3 = $ _____ $9 \times 2 = $ _____ $7 \times 3 = $ _____

C. $7 \times 2 = $ _____ $6 \times 3 = $ _____ $5 \times 2 = $ _____

 When multiplying by 0, the product is always 0. When multiplying by 1, the product is always the other factor.

D.

$$\begin{array}{r} 1 \\ \times\ 2 \\ \hline \end{array} \qquad \begin{array}{r} 8 \\ \times\ 3 \\ \hline \end{array} \qquad \begin{array}{r} 2 \\ \times\ 5 \\ \hline \end{array} \qquad \begin{array}{r} 0 \\ \times\ 3 \\ \hline \end{array} \qquad \begin{array}{r} 3 \\ \times\ 2 \\ \hline \end{array} \qquad \begin{array}{r} 2 \\ \times\ 7 \\ \hline \end{array}$$

E.

$$\begin{array}{r} 4 \\ \times\ 2 \\ \hline \end{array} \qquad \begin{array}{r} 3 \\ \times\ 3 \\ \hline \end{array} \qquad \begin{array}{r} 1 \\ \times\ 3 \\ \hline \end{array} \qquad \begin{array}{r} 6 \\ \times\ 2 \\ \hline \end{array} \qquad \begin{array}{r} 0 \\ \times\ 2 \\ \hline \end{array} \qquad \begin{array}{r} 3 \\ \times\ 1 \\ \hline \end{array}$$

Picture Perfect

 *An **array** shows a multiplication sentence. The first factor tells how many rows there are. The second factor tells how many are in each row. Here is an array for the multiplication sentence 4 x 4 = 16.*

$$\begin{array}{r} \textbf{4 rows} \\ \times \quad \textbf{4 rows} \\ \hline \textbf{16 in all} \end{array}$$

Solve each problem by creating an array.

A. 3 x 4 =	**B.** 6 x 5 =
C. 2 x 5 =	**D.** 6 x 4 =
E. 8 x 4 =	**F.** 3 x 5 =

Shining Brightly

Multiply. Then write the letter of the problem that matches each product below to learn the names of two of the brightest stars.

B. 3
x 4

R. 1
x 4

A. 2
x 4

F. 8
x 4

P. 7
x 4

S. 6
x 5

U. 3
x 5

E. 1
x 5

U. 4
x 4

I. 5
x 5

G. 0
x 5

S. 2
x 5

O. 4 x 5 = _____

D. 9 x 5 = _____

I. 9 x 4 = _____

N. 6 x 4 = _____

S. 7 x 5 = _____

C. 5 x 8 = _____

Two of the brightest stars are

____ ____ ____ ____ ____ ____ **and** ____ ____ ____ ____ ____ ____ ____ .
10   25   4   36   16   30   40   8   24   20   28   15   35

A Wheel of Facts

Multiply each number in the center by the numbers on the tire. Write your answers inside the wheel.

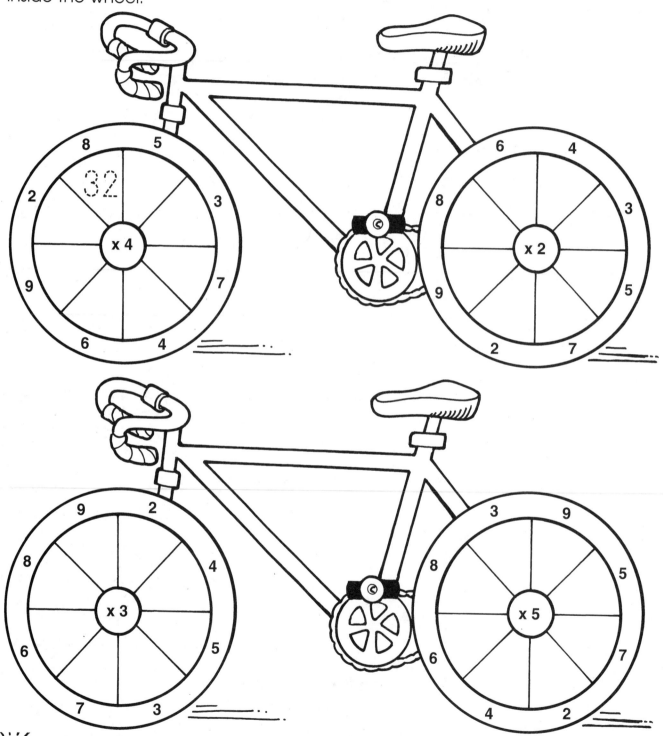

The bike team has 4 members. Each biker rides 9 miles every day. How many miles does the team ride every day altogether?

Product Drop

Multiply.

4
x 6

5 x 2 = _____

$$\begin{array}{r} 2 \\ x\ 6 \\ \hline \end{array}$$

$$\begin{array}{r} 3 \\ x\ 7 \\ \hline \end{array}$$

$$\begin{array}{r} 7 \\ x\ 3 \\ \hline \end{array}$$

$$\begin{array}{r} 7 \\ x\ 1 \\ \hline \end{array}$$

6 x 8 = _____

$$\begin{array}{r} 3 \\ x\ 6 \\ \hline \end{array}$$

$$\begin{array}{r} 5 \\ x\ 6 \\ \hline \end{array}$$

$$\begin{array}{r} 6 \\ x\ 5 \\ \hline \end{array}$$

$$\begin{array}{r} 4 \\ x\ 7 \\ \hline \end{array}$$

7 x 5 = _____

$$\begin{array}{r} 3 \\ x\ 7 \\ \hline \end{array}$$

6 x 6 = _____

$$\begin{array}{r} 9 \\ x\ 7 \\ \hline \end{array}$$

$$\begin{array}{r} 6 \\ x\ 7 \\ \hline \end{array}$$

$$\begin{array}{r} 6 \\ x\ 4 \\ \hline \end{array}$$

$$\begin{array}{r} 3 \\ x\ 7 \\ \hline \end{array}$$

$$\begin{array}{r} 7 \\ x\ 9 \\ \hline \end{array}$$

$$\begin{array}{r} 7 \\ x\ 7 \\ \hline \end{array}$$

9 x 6 = _____

$$\begin{array}{r} 8 \\ x\ 6 \\ \hline \end{array}$$

$$\begin{array}{r} 7 \\ x\ 6 \\ \hline \end{array}$$

8 x 7 = _____

5 x 6 = _____

7 x 4 = _____

Color by using the following product code.

0–10 = purple   21–30 = blue   41–50 = yellow   61–70 = pink

11–20 = orange   31–40 = red   51–60 = green

Sweet Success

Multiply.

A. 6 x 6 = _____   2 x 7 = _____

B. 1 x 7 = _____   5 x 6 = _____

C. 2 x 6 = _____   4 x 7 = _____

D. 0 x 7 = _____   7 x 7 = _____

E.
```
     6        2        9        4        6        3
   x 7      x 6      x 7      x 6      x 6      x 7
```

F.
```
     3        8        1        5        9        7
   x 6      x 7      x 6      x 7      x 6      x 6
```

 Ashley bought 4 flowers to plant in each pot. She has 7 pots. How many flowers did she buy in all? Using the pots below, draw a model to solve the problem. Then write a number sentence.

Name _____

The Product Trail

Multiply to get the lion back to its little cub.

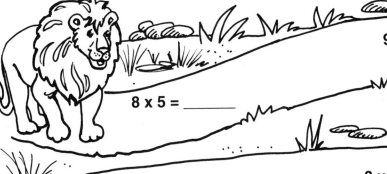

9 x 8 = _____

8 x 5 = _____

9
x 2
____

3 x 8 = _____

9 x 4 = _____

6
x 8
____

8 x 1 = _____   9 x 9 = _____   8 x 4 = _____

8
x 7
____

9 x 3 = _____   8 x 8 = _____   6 x 9 = _____

8
x 2
____

9 x 0 = _____

5 x 9 = _____

  There are 8 lions in the jungle. Each has 2 cubs. How many cubs are there altogether?

Climbing to the Top

Multiply.

A.   9   8   8   8   8
      x 6   x 9   x 5   x 6   x 3

B.   9   9   7   2   4
      x 3   x 9   x 8   x 9   x 8

C.   9   9   2   8   6
      x 8   x 0   x 8   x 8   x 9

D.   9   9   1   8   0
      x 4   x 7   x 9   x 4   x 8

E.   3   5   7   1   5
      x 9   x 8   x 9   x 8   x 9

 Circle the problem in Row E with the same product as 2 x 4.
Circle the problem in Row D with the same product as 3 x 3.
Circle the problem in Row C with the same product as 4 x 4.
Circle the problem in Row B with the same product as 6 x 3.
Circle the problem in Row A with the same product as 4 x 6.
Did you find your way to the top?

Name _____

The Number Man

Multiply two factors in the triangles. Write each product in the circle between the two factors.

On another sheet of paper, write two word problems about the Number Man using the multiplication facts 4 x 4 and 7 x 3.

The Case of the Missing Factors

Complete the multiplication chart.

x	0	1	2	3	4	5	6	7	8	9
0				0						
1										
2		2								
3										
4										
5										
6										
7								49		
8										
9										

Using the chart, help Detective Dan find each missing factor.

A. 4 x _____ = 12   7 x _____ = 14   3 x _____ = 27

B. 5 x _____ = 30   6 x _____ = 36   8 x _____ = 64

C. _____ x 4 = 36   _____ x 3 = 24   _____ x 9 = 18

D. _____ x 8 = 56   _____ x 9 = 81   _____ x 1 = 6

On another sheet of paper, write five missing factor number sentences. Have a friend solve them. Check your friend's work.

Name _____

Triple Hit

Multiply around the bases. Start at first base on each baseball field. Then multiply the number on each base in order. Write each product on home plate.

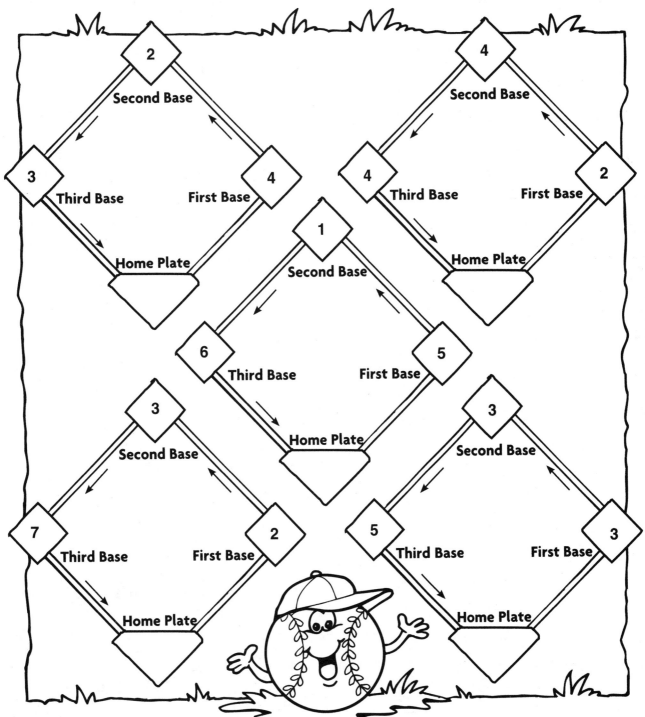

💡 **Four players each have 2 boxes of balls. There are 4 balls in each box. How many balls do the players have altogether?**

Name _____

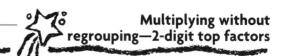

One Step, Two Step

 To multiply a two-digit number by a one-digit number, follow these steps.

1. Multiply the ones digit.

$4 \times 2 = 8$
```
  34
x  2
-----
   8
```

2. Multiply the tens digit.

$3 \times 2 = 6$
```
  34
x  2
-----
  68
```

Multiply. Then use the code to answer the riddle below.

U.
```
  23
x  2
```

A.
```
  12
x  3
```

Q.
```
  14
x  2
```

E.
```
  31
x  3
```

E.
```
  33
x  3
```

J.
```
  71
x  3
```

D.
```
  83
x  2
```

!
```
  22
x  4
```

C.
```
  24
x  2
```

N.
```
  11
x  5
```

A.
```
  43
x  3
```

S.
```
  74
x  2
```

What kind of dancers are math teachers?

R.
```
  52
x  4
```

S.
```
  33
x  2
```

| 66 | 28 | 46 | 129 | 208 | 99 |

| 166 | 36 | 55 | 48 | 93 | 208 | 148 | 88 |

Name _____

Times Race

Multiply. Time yourself to see how fast you can finish the race.

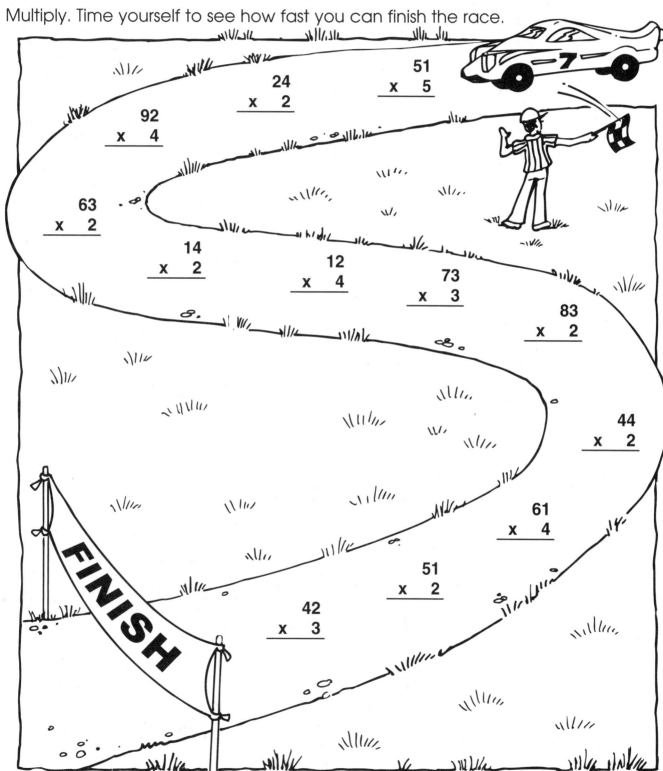

$$\begin{array}{r} 51 \\ \times\ 5 \\ \hline \end{array}$$

$$\begin{array}{r} 24 \\ \times\ 2 \\ \hline \end{array}$$

$$\begin{array}{r} 92 \\ \times\ 4 \\ \hline \end{array}$$

$$\begin{array}{r} 63 \\ \times\ 2 \\ \hline \end{array}$$

$$\begin{array}{r} 14 \\ \times\ 2 \\ \hline \end{array}$$

$$\begin{array}{r} 12 \\ \times\ 4 \\ \hline \end{array}$$

$$\begin{array}{r} 73 \\ \times\ 3 \\ \hline \end{array}$$

$$\begin{array}{r} 83 \\ \times\ 2 \\ \hline \end{array}$$

$$\begin{array}{r} 44 \\ \times\ 2 \\ \hline \end{array}$$

$$\begin{array}{r} 61 \\ \times\ 4 \\ \hline \end{array}$$

$$\begin{array}{r} 51 \\ \times\ 2 \\ \hline \end{array}$$

$$\begin{array}{r} 42 \\ \times\ 3 \\ \hline \end{array}$$

FINISH

**Three race cars raced around the track. Each race car completed 32 laps. How many
laps in all did the race cars complete? Solve the problem on another sheet of paper.**

Carry Carefully

 *Sometimes regrouping will be needed when
multiplying with a two-digit number. Follow these
steps to solve the problem.*

1. Multiply the ones.
   Regroup if needed.

   $7 \times 6 = 42$

   $$\begin{array}{r} 4 \\ 47 \\ \times \quad 6 \\ \hline 2 \end{array}$$

2. Multiply the tens.
   Add the extra tens.

   $4 \times 6 = 24$

   $24 + 4 = 28$

   $$\begin{array}{r} 4 \\ 47 \\ \times \quad 6 \\ \hline 282 \end{array}$$

Multiply. Remember to regroup.

A. $\begin{array}{r} \square \\ 36 \\ \times \quad 4 \\ \hline \end{array}$ $\begin{array}{r} \square \\ 25 \\ \times \quad 5 \\ \hline \end{array}$ $\begin{array}{r} \square \\ 63 \\ \times \quad 7 \\ \hline \end{array}$

B. $\begin{array}{r} \square \\ 83 \\ \times \quad 8 \\ \hline \end{array}$ $\begin{array}{r} \square \\ 72 \\ \times \quad 6 \\ \hline \end{array}$ $\begin{array}{r} \square \\ 29 \\ \times \quad 4 \\ \hline \end{array}$ $\begin{array}{r} \square \\ 47 \\ \times \quad 6 \\ \hline \end{array}$ $\begin{array}{r} \square \\ 55 \\ \times \quad 7 \\ \hline \end{array}$

C. $\begin{array}{r} \square \\ 62 \\ \times \quad 5 \\ \hline \end{array}$ $\begin{array}{r} \square \\ 96 \\ \times \quad 2 \\ \hline \end{array}$ $\begin{array}{r} \square \\ 58 \\ \times \quad 5 \\ \hline \end{array}$ $\begin{array}{r} \square \\ 49 \\ \times \quad 3 \\ \hline \end{array}$ $\begin{array}{r} \square \\ 96 \\ \times \quad 4 \\ \hline \end{array}$

       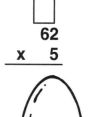 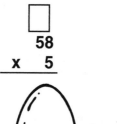 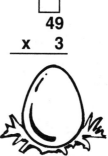 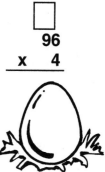

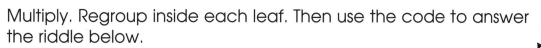
Falling for Multiplication

Multiply. Regroup inside each leaf. Then use the code to answer the riddle below.

N.   34
      x   6

C.   17
      x   3

A.   46
      x   4

H.   62
      x   5

I.   53
      x   6

B.   72
      x   7

S.   28
      x   4

K.   48
      x   2

R.   18
      x   6

B.   39
      x   3

Where does a tree keep its money?

$\overline{318}$ $\overline{204}$ $\overline{117}$ $\overline{108}$ $\overline{184}$ $\overline{204}$ $\overline{51}$ $\overline{310}$ $\overline{504}$ $\overline{184}$ $\overline{204}$ $\overline{96}$ $\underline{\quad}$!

Name _____

Let's Review!

Multiply. Remember to regroup if needed.

A.
$$\begin{array}{r} 53 \\ \times\ 3 \\ \hline \end{array} \qquad \begin{array}{r} 63 \\ \times\ 2 \\ \hline \end{array} \qquad \begin{array}{r} 46 \\ \times\ 4 \\ \hline \end{array} \qquad \begin{array}{r} 73 \\ \times\ 4 \\ \hline \end{array}$$

B.
$$\begin{array}{r} 34 \\ \times\ 6 \\ \hline \end{array} \qquad \begin{array}{r} 82 \\ \times\ 4 \\ \hline \end{array} \qquad \begin{array}{r} 35 \\ \times\ 5 \\ \hline \end{array} \qquad \begin{array}{r} 27 \\ \times\ 4 \\ \hline \end{array}$$

C.
$$\begin{array}{r} 75 \\ \times\ 2 \\ \hline \end{array} \qquad \begin{array}{r} 23 \\ \times\ 7 \\ \hline \end{array} \qquad \begin{array}{r} 52 \\ \times\ 3 \\ \hline \end{array} \qquad \begin{array}{r} 32 \\ \times\ 2 \\ \hline \end{array}$$

D.
$$\begin{array}{r} 29 \\ \times\ 2 \\ \hline \end{array} \qquad \begin{array}{r} 38 \\ \times\ 5 \\ \hline \end{array} \qquad \begin{array}{r} 48 \\ \times\ 6 \\ \hline \end{array} \qquad \begin{array}{r} 84 \\ \times\ 2 \\ \hline \end{array}$$

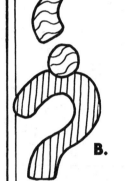

Circle each problem above that did not need regrouping. Is there a pattern?

Snack Time!

hot dog 72¢

popcorn 29¢

soda 95¢

pretzel 68¢

cotton candy 87¢

snow cone 43¢

Use the chart above to write a multiplication sentence for each problem. Multiply.

A. What is the cost of 4 hot dogs?

_____ X _____ = $_____

B. How much will 6 popcorn bags cost?

_____ X _____ = $_____

C. If you buy 3 pretzels, how much will you spend?

_____ X _____ = $_____

D. What is the cost of 3 cotton candies?

_____ X _____ = $_____

E. What is the cost of 2 snow cones and 2 popcorns?

_____ X _____ = $_____

_____ X _____ = $_____

_____ + _____ = $_____

F. You bought 3 sodas and 2 pretzels. How much did you spend?

_____ X _____ = $_____

_____ X _____ = $_____

_____ + _____ = $_____

 On another sheet of paper, write a multiplication word problem using the prices of the snacks above. Solve.

Sweet and Juicy

Multiply. Color an apple if you find its product.

A.

203	411	310	212
x 3	x 2	x 1	x 3

B.

110	141	130	302
x 7	x 2	x 3	x 3

C.

114	524	333	230
x 2	x 1	x 3	x 2

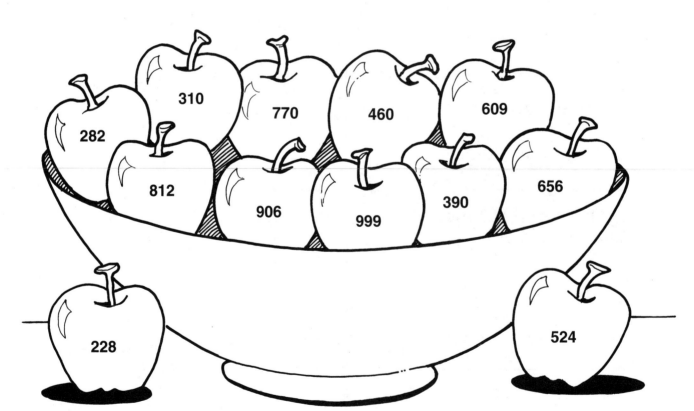

282 310 770 460 609
812 906 999 390 656
228 524

 There are 4 crates of apples to unpack. If each crate has 102 apples, how many apples are there altogether?

Zany and Brainy

Multiply.

A. 314   230
     x 2   x 3

B. 432   521
     x 3   x 4

C. 604   702
     x 2   x 3

D. 723   921   241   813
     x 3   x 2   x 2   x 3

E. 112   124   303   620
     x 3   x 2   x 3   x 4

 The school science lab has 3 sets of test tubes. Each set has 112 tubes. How many are there in all?

Name _____

Check It Out!

Multiply. Then use a calculator to check your work.

A.   315
       x   9

B.   456
       x   4

C.   675
       x   5

D.   764
       x   7

E.   219
       x   8

F.   968
       x   3

G.   391
       x   4

H.   532
       x   6

I.   808
       x   4

J.   270
       x   9

 On another sheet of paper, write five multiplication problems with a three-digit number. Multiply. Check each answer with a calculator.

Name _____

Number Fun With Barky

Write a number sentence for each problem. Solve.

A. Connor's dog, Barky, made 3 holes in the backyard. Connor's dad had to fill each hole with 78 scoops of dirt. How many scoops did his dad need in all?

B. Barky got into Steve's closet. He chewed up 8 pairs of shoes. How many shoes did he chew altogether?

C. Adrienne went to the store to buy doggie treats. She bought 6 boxes of doggie treats. Each box has 48 treats. How many treats in all did Adrienne buy?

D. Terri took Barky to the vet for 3 shots. Each shot cost $2.65. How much money did Terri pay the vet?

E. Max's job is to keep Barky's water bowl full. If he fills it 3 times a day for 24 days, how many times did he fill the bowl altogether?

F. Barky runs around the block 4 times every day. How many times does he run around the block in 5 days?

 On another sheet of paper, write your own Barky word problem. Solve.

Tic-Tac-Toe

Multiply. If the ones digit in the product is less than five, mark an *O* in the box.
If the ones digit is five or greater, mark an *X*. Are there three in a row?

Game 1

$ .56 x 7	129 x 8	42 x 3
238 x 3	251 x 6	$1.32 x 4
62 x 4	83 x 4	$1.85 x 2

Game 2

97 x 3	$1.89 x 4	224 x 4
$ .76 x 4	55 x 8	252 x 3
225 x 4	304 x 2	58 x 3

On another sheet of paper, make up your own multiplication tic-tac-toe game.

Working With Area

 *The **area** of an object is the number of square units needed to cover its surface. To find the area, multiply the length by the width .*

4 x 2 = 8 square units

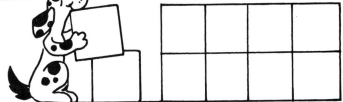

Write a multiplication sentence to figure the area of each object. Multiply.

A.

A = _____
square units

B.

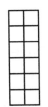

A = _____
square units

C.

A = _____
square units

D.

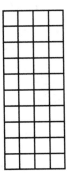

A = _____
square units

E.

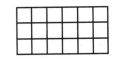

A = _____
square units

F.

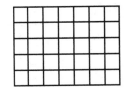

A = _____
square units

G.

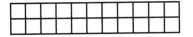

A = _____
square units

H.

A = _____
square units

I.

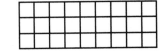

A = _____
square units

 The playground at school is 36 yards long and 9 yards wide. What is the area of the playground?

What Is Division?

 To divide means to make equal groups. The total number being divided is called the **dividend**. *The number of groups the total is to be divided into is called the* **divisor**. *The answer is called the* **quotient**. $6 ÷ 2 = 3$

total number (dividend–6)	number of groups (divisor–2)	number in each group (quotient–3)
		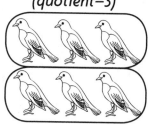

The Bird House has 10 birds in all. The zookeeper wants to put the birds into the 5 new cages he bought. How many birds will he put in each cage?

Solve this problem by drawing a picture. Draw the number of birds you think need to go in each cage. (Hint: Each cage must have the same number of birds.) Then complete the number sentence.

Total Number of Birds	Number of Cages	Number of Birds in Each Cage
_____ ÷	5 =	_____

What if the zookeeper only had 2 cages? How many birds would go in each cage? Draw a picture. Then write a number sentence.

_____ ÷ _____ = _____

Name _____

Filled With Stars

Draw a circle around the correct number of stars to show each division problem. Complete each number sentence.

A. $8 \div 2 =$ __4__

B. $6 \div 3 =$ __2__

C. $12 \div 3 =$ __4__

D. $10 \div 2 =$ __5__

E. $18 \div 3 =$ __6__

F. $9 \div 3 =$ __3__

G. $16 \div 2 =$ __8__

H. $15 \div 3 =$ __5__

 On another sheet of paper, write a number sentence and draw a picture to show 12 stars divided into 2 groups.

Alien Adventure

Divide.

A. $6 \div 2 = \underline{3}$ $9 \div 3 = \underline{3}$ $10 \div 2 = \underline{5}$

B. $12 \div 3 = \underline{4}$ $14 \div 2 = \underline{7}$ $8 \div 2 = \underline{4}$

C. $2 \div 2 = \underline{1}$ $18 \div 3 = \underline{6}$ $24 \div 3 = \underline{8}$

D. $2\overline{)12}$ $3\overline{)21}$

E. $3\overline{)6}$ $3\overline{)3}$

F. $3\overline{)15}$ $2\overline{)16}$

G. There are 18 aliens ready to board their spaceships. If 6 aliens get on each spaceship, how many spaceships do they need? Draw a picture to show the problem. Then write a number sentence to solve the problem.

2 spaceship

 On another sheet of paper, using the numbers 12 and 3, write your own word problem. Draw a picture and write a number sentence. Solve.

Name _____

Flying Back

 You can use a number line to help divide. Count back in equal groups to 0.

$28 \div 4 = 7$

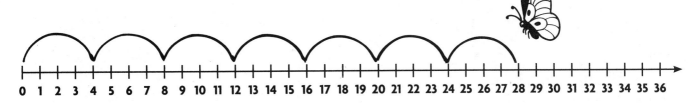

Divide. Use the number line to help you.

A. $4\overline{)16}$ $4\overline{)36}$ $4\overline{)4}$ $4\overline{)24}$

B. $4\overline{)20}$ $4\overline{)12}$ $4\overline{)32}$ $4\overline{)8}$

C. $32 \div 4 = \underline{8}$ $16 \div 4 = \underline{4}$ $20 \div 4 = \underline{5}$

Divide. Use the number line to help you.

D. $5\overline{)15}$ $5\overline{)5}$ $5\overline{)40}$ $5\overline{)45}$

E. $5\overline{)25}$ $5\overline{)10}$ $5\overline{)20}$ $5\overline{)30}$

Dividing Is a Breeze

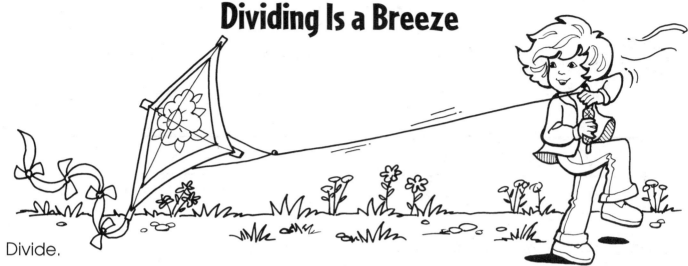

Divide.

A. 30 ÷ 5 = _____   32 ÷ 4 = _____   45 ÷ 5 = _____   5 ÷ 5 = _____

B. 36 ÷ 4 = _____   20 ÷ 4 = _____   25 ÷ 5 = _____   28 ÷ 4 = _____

C. 5$\overline{)10}$ 4$\overline{)16}$ 5$\overline{)40}$ 5$\overline{)45}$ 4$\overline{)20}$

D. 4$\overline{)12}$ 5$\overline{)35}$ 4$\overline{)8}$ 5$\overline{)15}$ 4$\overline{)24}$

E. Lisa tied a total of 12 ribbons on her kites. If she tied 4 ribbons on each kite, how many kites does Lisa have?

 There were 36 people flying kites in the park. There were an equal number of yellow, green, orange, and blue kites. How many kites are there of each color?

Dividing Evenly

 *If a number can be evenly divided by a number, it is **divisible** by that number. For example*

2, 4, 6, 8, 10, and 12 are all divisible by 2.
3, 6, 9, 12, 15, and 18 are all divisible by 3.

Look at each number on the chart. Decide if the number is divisible by 2, 3, 4, or 5. Circle each number using the color code. Some numbers will be circled more than once.

Numbers Divisible by	Circle
2	red
3	green
4	yellow
5	blue

4	7	9	21	15	22
25	28	40	18	30	5
31	32	11	14	3	35
2	27	12	36	16	20
8	6	24	29	10	45

 Choose two numbers you circled with three different colors. On another sheet of paper, write the three division number sentences for each number.

Leaping Lily Pads

Divide.

A. $6\overline{)30}$ $7\overline{)28}$ $6\overline{)12}$ $6\overline{)48}$

B. $7\overline{)49}$ $6\overline{)18}$ $7\overline{)35}$ $6\overline{)6}$

C. $21 \div 7 =$ _____ $7 \div 7 =$ _____ $42 \div 6 =$ _____

D. $14 \div 7 =$ _____ $24 \div 6 =$ _____ $36 \div 6 =$ _____

Divide by 6.

0	6	12	18	24	30	36	42	48	54

Divide by 7.

0	7	14	21	28	35	42	49	56	63

 Circle the division number sentence above that shows the picture of the frogs on the lily pads. On another sheet of paper, write a word problem using another number sentence from above.

Name _____

Field Trip Fun

Divide.

A. 42 ÷ 7 = _____   54 ÷ 6 = _____   36 ÷ 6 = _____

B. 24 ÷ 6 = _____   63 ÷ 7 = _____   48 ÷ 6 = _____

C. 14 ÷ 7 = _____   56 ÷ 7 = _____   28 ÷ 7 = _____

D. 49 ÷ 7 = _____   60 ÷ 6 = _____   42 ÷ 6 = _____

E. Fifty-six students went on a field trip to the zoo. They traveled in 7 vans. How many students were in each van?

F. When the students went to the monkey house, they found it was divided into 6 rooms. The same number of monkeys were in each room. There were 24 monkeys in all. How many monkeys were in each room?

 Add the quotients in each row of problems above. Which row has a sum equal to the total number of monkeys at the Monkey House?

Name _____

The Skate Divide

Color each skate and helmet with the correct quotient.

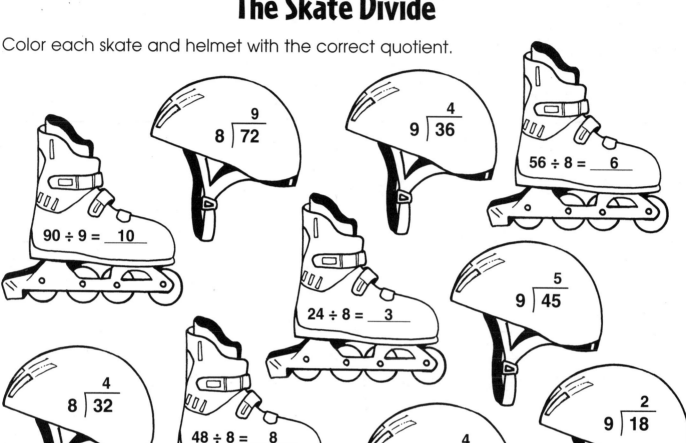

$8\overline{)72}$ → 9

$9\overline{)36}$ → 4

$56 \div 8 = \underline{6}$

$90 \div 9 = \underline{10}$

$24 \div 8 = \underline{3}$

$9\overline{)45}$ → 5

$8\overline{)32}$ → 4

$48 \div 8 = \underline{8}$

$8\overline{)40}$ → 4

$9\overline{)18}$ → 2

$63 \div 9 = \underline{7}$

$8\overline{)64}$ → 8

$80 \div 8 = \underline{11}$

$81 \div 9 = \underline{9}$

Peter wants to skate 18 miles in the next 9 days. If he skates an equal number of miles each day, how many miles will he need to skate each day?

On another sheet of paper, write the correct division number sentences for the problems above that you did not color.

Name _____

Dividing Race

Use a stopwatch to time how long it takes to divide each runner's path to the finish line.

$81 \div 9 =$ _____

$16 \div 8 =$ _____

$32 \div 8 =$ _____

$45 \div 9 =$ _____

$8 \overline{)40}$

$9 \overline{)63}$

$64 \div 8 =$ _____

$18 \div 9 =$ _____

$27 \div 9 =$ _____

$9 \div 9 =$ _____

$9 \overline{)36}$

$8 \overline{)24}$

FINISH

$8 \div 0 =$ _____

$56 \div 8 =$ _____

$9 \overline{)54}$

$8 \overline{)72}$

$48 \div 8 =$ _____

$80 \div 8 =$ _____

 Last week in track practice, Andy ran 36 miles. He ran the same number of miles on each of the 4 days. How many miles did he run each day?

Fishy Fact Families

 Division is the opposite of multiplication. The dividend, divisor, and quotient can be used to write multiplication sentences. The division and multiplication sentences are called a **fact family.**

15 ÷ 3 = 5 (15 divided into 3 equal groups)
15 ÷ 5 = 3 (15 divided into 5 equal groups)
3 x 5 = 15 (3 groups of 5)
5 x 3 = 15 (5 groups of 3)

Use the numbers from each fish family
to write fact family number sentences.

A.

____ X ____ = ____
____ X ____ = ____
____ ÷ ____ = ____
____ ÷ ____ = ____

(fish: 4, 3, 12)

B.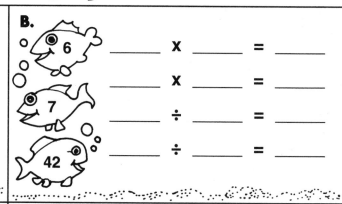

____ X ____ = ____
____ X ____ = ____
____ ÷ ____ = ____
____ ÷ ____ = ____

(fish: 6, 7, 42)

C.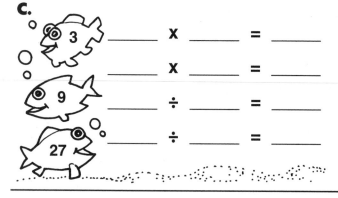

____ X ____ = ____
____ X ____ = ____
____ ÷ ____ = ____
____ ÷ ____ = ____

(fish: 3, 9, 27)

D.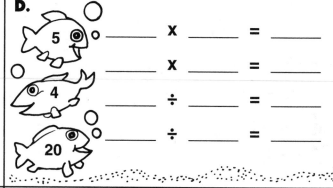

____ X ____ = ____
____ X ____ = ____
____ ÷ ____ = ____
____ ÷ ____ = ____

(fish: 5, 4, 20)

E.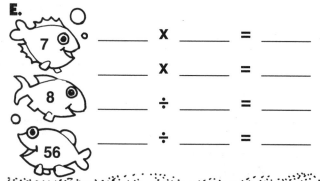

____ X ____ = ____
____ X ____ = ____
____ ÷ ____ = ____
____ ÷ ____ = ____

(fish: 7, 8, 56)

F.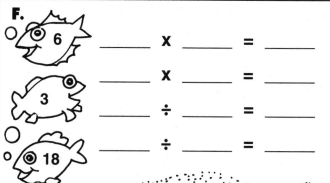

____ X ____ = ____
____ X ____ = ____
____ ÷ ____ = ____
____ ÷ ____ = ____

(fish: 6, 3, 18)

Over the Hurdles

 Sometimes when you try to divide a number into equal groups, part of the number is left over. This is called the **remainder**. Use these steps to find the remainder.

1.
$$5 \overline{)\ 16\ }$$

Think: 5 x ___ is the closest to 16?

2.
$$\begin{array}{r} 3 \\ 5 \overline{)\ 16\ } \\ -15 \\ \hline 1 \end{array}$$

3.
$$\begin{array}{r} 3\ R\ 1 \\ 5 \overline{)\ 16\ } \\ -15 \\ \hline 1 \end{array}$$

There are 5 groups of 3 with 1 left over.

Divide.

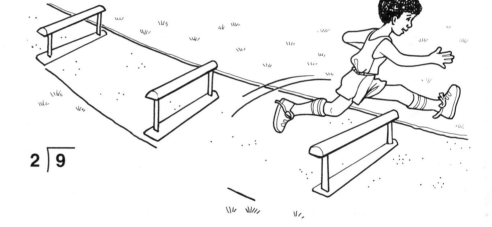

A.
$$\begin{array}{r} 1\ R4 \\ 6 \overline{)\ 10\ } \\ -6 \\ \hline 4 \end{array}$$

$$2 \overline{)\ 9\ }$$

B.
$$3 \overline{)\ 20\ } \qquad 2 \overline{)\ 19\ } \qquad 6 \overline{)\ 47\ } \qquad 6 \overline{)\ 41\ }$$

C.
$$7 \overline{)\ 51\ } \qquad 2 \overline{)\ 15\ } \qquad 3 \overline{)\ 22\ } \qquad 7 \overline{)\ 48\ }$$

D.
$$2 \overline{)\ 11\ } \qquad 4 \overline{)\ 26\ } \qquad 6 \overline{)\ 19\ } \qquad 5 \overline{)\ 27\ }$$

 Louie jumps hurdles for his track meet. He jumps the same number in each race. In his last 6 races, he jumped 36 hurdles. On another sheet of paper, draw a picture to show the problem.

More Remainders

Divide.

A. $5\overline{)41}$ $6\overline{)52}$ $3\overline{)19}$ $8\overline{)74}$

B. $4\overline{)29}$ $2\overline{)13}$ $7\overline{)38}$ $9\overline{)46}$

C. $5\overline{)21}$ $6\overline{)31}$ $3\overline{)26}$ $8\overline{)57}$

D. $4\overline{)14}$ $2\overline{)7}$ $7\overline{)65}$ $9\overline{)51}$

E. $3\overline{)13}$ $6\overline{)39}$ $5\overline{)14}$ $8\overline{)50}$

 Candy's mom bought 56 apples to make 8 pies. If she used an equal number of apples in each pie, how many apples did she use in each pie? Solve on another sheet of paper.

Name _____

Figure It Out

 *Remember: The **quotient** tells how many equal groups you can make. The **remainder** tells how many are left over.*

Divide. Answer each question.

A. A clothing store clerk has 14 sweaters. He wants to put them in equal stacks on 3 shelves. How many sweaters will be in each stack?	**B.** Mary has 57¢. She wants to buy candy canes that cost 9¢ each. How many candy canes can she buy?
C. Rosa needs to bake 71 cookies. Each cookie sheet holds 8 cookies. How many cookies are on the unfilled cookie sheet?	**D.** There are 17 cars waiting to be parked. There are an equal number of parking spots on 3 different levels. How many cars will not find a parking spot?
E. Luis is putting 74 cans into cartons. Each carton holds 8 cans. How many cans will be in the unfilled carton?	**F.** Don bought 85 crates of flowers. He separates them into groups of 9. How many equal groups did he have?

 On another sheet of paper, write a word problem in which the quotient will be the answer. Write another word problem in which the remainder will be the answer.

Keep on Dividing

 Use these steps when dividing with greater dividends.

1. *Divide the tens digit in the dividend by the divisor. Write the answer above the tens digit.*

$$\begin{array}{r} 2 \\ 4\,\overline{)\,84} \end{array}$$

2. *Multiply the partial quotient by the divisor. Write the answer below the tens digit. Subtract. Bring down the ones digit.*

$$\begin{array}{r} 2 \\ 4\,\overline{)\,84} \\ -8\downarrow \\ \hline 04 \end{array}$$

3. *Divide the ones digit by the divisor. Write the answer above the ones digit. Multiply. Subtract.*

$$\begin{array}{r} 21 \\ 4\,\overline{)\,84} \\ -8\downarrow \\ \hline 04 \\ -4 \\ \hline 0 \end{array}$$

Divide.

A.

$3\,\overline{)\,66}$ $\qquad$ $2\,\overline{)\,48}$ $\qquad$ $3\,\overline{)\,93}$ $\qquad$ $3\,\overline{)\,39}$

B.

$3\,\overline{)\,96}$ $\qquad$ $3\,\overline{)\,63}$ $\qquad$ $2\,\overline{)\,68}$ $\qquad$ $9\,\overline{)\,90}$

C.

$3\,\overline{)\,99}$ $\qquad$ $3\,\overline{)\,69}$ $\qquad$ $2\,\overline{)\,80}$ $\qquad$ $5\,\overline{)\,55}$

Name _____

Dividing the Loot

 Remember to follow each step when dividing larger numbers.

1. Divide the tens digit
   by the divisor.
   Multiply. Subtract.

```
     1
 3 ) 45
    -3
     1
```

2. Bring down the ones
   digit. Divide this
   number by the divisor.

```
     1 5
 3 ) 45
    -3↓
     1 5
```

3. Multiply. Subtract.

```
     1 5
 3 ) 45
    -3↓
     1 5
    -1 5
       0
```

Divide.

A.

2) 58   5) 85   6) 72   5) 90

B.

3) 48   8) 96   2) 74   4) 92

C.

6) 78   4) 76   5) 65   4) 60

 **Andrew has 87 marbles. He divides them into 3 bags.
How many marbles are in each bag? Solve. Then
circle the problem above with the same quotient.**

It's All Relative

➡️ *Remember that multiplication and division are related. Multiplying the quotient by the divisor will tell you the dividend.*

Hi! Aren't we related?

You bet! When you multiply us, our missing product is the missing dividend!

$? \div 7 = 8$

Write each missing dividend.

A. _____ $\div 9 = 7$ _____ $\div 4 = 6$ _____ $\div 6 = 6$ _____ $\div 5 = 7$

B. _____ $\div 3 = 3$ _____ $\div 2 = 9$ _____ $\div 8 = 6$ _____ $\div 9 = 9$

C. _____ $\div 4 = 8$ _____ $\div 3 = 7$ _____ $\div 2 = 8$ _____ $\div 6 = 3$

D. _____ $\div 8 = 8$ _____ $\div 1 = 9$ _____ $\div 5 = 6$ _____ $\div 7 = 1$

Are we missing?

E. _____ $\div 4 = 40$ _____ $\div 3 = 30$ _____ $\div 3 = 100$

F. _____ $\div 7 = 60$ _____ $\div 5 = 60$ _____ $\div 2 = 40$

 At our family reunion picnic, 8 people sat at each picnic table. We needed 16 tables. How many people altogether were at the reunion?

Name _____

Division Treats

candy cane
5¢

licorice
7¢

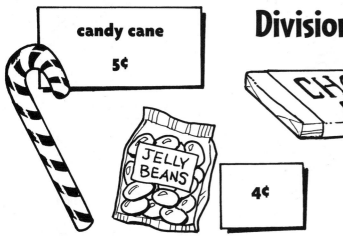

JELLY
BEANS

4¢

9¢

cookies
6¢ each

Write a number sentence for each problem. Solve.

A. Suzanne has 96¢. How many cookies can she buy?	**B.** Lee has 98¢. How many candy bars can she buy? How much money will she have left over?
C. Jose has 72¢. How many bags of jelly beans can he buy?	**D.** Connie has 84¢. How many cookies can she buy?
E. Toby is in the mood for candy canes. How many can he buy with 63¢?	**F.** Ann is buying licorice for her friends. How many pieces can she buy for 74¢? What could she buy with the remaining money?

 The price of which items above can be evenly divided into $1.00?

Decision Time

Decide whether to multiply or
divide. Solve.

A. Ellen baked 75 cookies in 3 hours.
Joe baked 96 cookies in 4 hours.
Who baked the most cookies per
hour?

B. James pitched 18 times in each
inning of the ball game. How many
times did he pitch in the 9 innings?

C. Lana bought 4 20-ounce sodas.
How many 4-ounce servings can
she give her party guests?

D. Cory's mom sent him to the store
for eggs. He bought 4 cartons of a
dozen eggs. How many eggs did
he purchase in all?

E. Maria made bracelets for her
friends. She put 9 beads on each.
She had 81 beads. How many
bracelets did she make?

F. It costs 50¢ per hour to park at
the beach. How much did it cost
David's parents to park for 8 hours?

 **On another sheet of paper, write a sentence that explains how you know whether to
multiply or divide.**

Page 4
Check students' models.
A. 3 x 3 = 9 points; B. 4 x
2 = 8 hockey sticks

Page 5
A. 4, 9, 12; B. 12, 18, 21;
C. 14, 18, 10; D. 2, 24, 10,
0, 6, 14; E. 8, 9, 3, 12, 0, 3

Page 6
Check students' arrays.
A. 12; B. 30; C. 10; D. 24;
E. 32; F. 15

Page 7
B. 12; R. 4; A. 8; F. 32;
P. 28; S. 30; U. 15; E. 5;
U. 16; I. 25; G. 0; S. 10;
O. 20; D. 45; I. 36; N. 24;
S. 35; C. 40; SIRIUS and
CANOPUS

Page 8

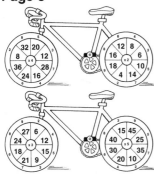

4 x 9 = 36 miles

Page 9

Check student's pictures.

Page 10
A. 36, 14; B. 7, 30; C. 12,
28; D. 0, 49; E. 42, 12,
63, 24, 36, 21; F. 18, 56,
6, 35, 54, 42; 4 x 7 = 28
flowers

Page 11

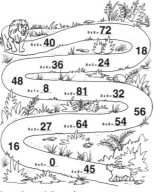

8 x 2 = 16 cubs

Page 12
A. 54, 72, 40, 48, 24;
B. 27, 81, 56, 18, 32;
C. 72, 0, 16, 64, 54; D. 36,
63, 9, 32, 0; E. 27, 40, 63,
8, 45; 1 x 8, 1 x 9, 2 x 8,
2 x 9, 8 x 3

Page 13

Page 14

x	0	1	2	3	4	5	6	7	8	9
0	0	0	0	0	0	0	0	0	0	0
1	0	1	2	3	4	5	6	7	8	9
2	0	2	4	6	8	10	12	14	16	18
3	0	3	6	9	12	15	18	21	24	27
4	0	4	8	12	16	20	24	28	32	36
5	0	5	10	15	20	25	30	35	40	45
6	0	6	12	18	24	30	36	42	48	54
7	0	7	14	21	28	35	42	49	56	63
8	0	8	16	24	32	40	48	56	64	72
9	0	9	18	27	36	45	54	63	72	81

A. 3, 2, 9; B. 6, 6, 8; C. 9,
8, 2; D. 7, 9, 6

Page 15
24, 32, 30, 42, 45

Page 16
U. 46; A. 36; Q. 28; E. 93;
E. 99; J. 213; D. 166; !
88; C. 48; N. 55; A. 129;
S. 148; R. 208; S. 66;
SQUARE DANCERS!

Page 17

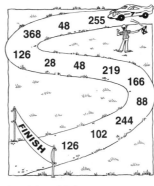

3 x 32 = 96 laps

Page 18
A. 144, 125, 441; B. 664,
432, 116, 282, 385;
C. 310, 192, 290, 147, 384

Page 19
N. 204; C. 51; A. 184;
H. 310; I. 318; B. 504;
S. 112; K. 96; R. 108;
B. 117;
IN BRANCH BANKS!

Page 20
A. 159, 126, 184, 292;
B. 204, 328, 175, 108;
C. 150, 161, 156, 64;
D. 58, 190, 288, 168; A. 53
x 3, 63 x 2; B. 82 x 4; C.
52 x 3, 32 x 2; D. 84 x 2;
2 problems in row A
1 problem in row B
2 problems in row C
1 problem in row D

Page 21
A. 72¢ X 4 = $2.88;
B. 29¢ X 6 = $1.74; C.
68¢ x 3 = $2.04; D. 87¢
x 3 = $2.61; E. 43¢ x 2 =
$.86, 29¢ x 2 = $.58, $.86
+ $.58 = $1.44; F. 95¢ x 3
= $2.85, 68¢ x 2 = $1.36,
$2.85 + $1.36 = $4.21

Page 22
A. 609, 822, 310, 636;
B. 770, 282, 390, 906;
C. 228, 524, 999, 460;
102 x 4 = 408 apples

Page 23
A. 628, 690; B. 1,296,
2,084; C. 1,208, 2,106;
D. 2,169, 1,842, 482,
2,439; E. 336, 248, 909,
2,480; 336 test tubes

Page 24
A. 2,835; B. 1,824;
C. 3,375; D. 5,348;
E. 1,752; F. 2,904;
G. 1,564; H. 3,192;
I. 3,232; J. 2,430

Page 25
A. 78 X 3 = 234 scoops of
dirt; B. 8 x 2 = 16 shoes;
C. 48 x 6 = 288 treats;
D. $2.65 x 3 = $7.95;
E. 24 x 3 = 72 times;
F. 4 x 5 = 20 times

Page 26
Game 1:

$3.92	1,032	126
O	O	X
714	1,506	$5.28
O	X	X
248	332	$3.70
X	O	O

Game 2:

291	$7.56	896
O	X	X
$3.04	440	756
O	O	X
900	608	174
O	X	O

Page 27
A. 4 x 3 = 12; B. 6 x 2 =
12; C. 4 x 6 = 24; D. 10 x
4 = 40; E. 3 x 6 = 18;
F. 5 x 7 = 35; G. 2 x 11 =
22; H. 5 x 4 = 20; I. 3 x 9
= 27; 36 x 9 = 324 square
yards

Page 28
Check students' drawings.
10 ÷ 5 = 2; 10 ÷ 2 = 5

Page 29
Check that students have circled the appropriate number of stars. A. 4; B. 2; C. 4; D. 5; E. 6; F. 3; G. 8; H. 5; 12 ÷ 2 = 6

Page 30
A. 3, 3, 5; B. 4, 7, 4; C. 1, 6, 8; D. 6, 7; E. 2, 1; F. 5, 8; G. 18 ÷ 6 = 3 spaceships

Page 31
A. 4, 9, 1, 6; B. 5, 3, 8, 2; C. 8, 4, 5; D. 3, 1, 8, 9; E. 5, 2, 4, 6

Page 32
A. 6, 8, 9, 1; B. 9, 5, 5, 7; C. 2, 4, 8, 9, 5; D. 3, 7, 2, 3, 6; E. 12 ÷ 4 = 3 kites; 36 ÷ 4 = 9 kites

Page 33
Check to see that students have circled the numbers with the appropriate colors. Divisible by 2: 4, 22, 28, 40, 18, 30, 32, 14, 2, 12, 36, 16, 20, 8, 6, 24, 10; Divisible by 3: 9, 21, 15, 18, 30, 3, 27, 12, 36, 6, 24, 45; Divisible by 4: 4, 28, 40, 32, 12, 36, 16, 20, 8, 24; Divisible by 5: 15, 25, 40, 30, 5, 35, 20, 10, 45

Page 34
A. 5, 4, 2, 8; B. 7, 3, 5, 1; C. 3, 1, 7; D. 2, 4, 6

0	6	12	18	24	30	36	42	48	54
0	1	2	3	4	5	6	7	8	9

0	7	14	21	28	35	42	49	56	63
0	1	2	3	4	5	6	7	8	9

Circle 30 ÷ 6 = 5.

Page 35
A. 6, 9, 6; B. 4, 9, 8; C. 2, 8, 4; D. 7, 10, 7; E. 56 ÷ 7 = 8 students; F. 24 ÷ 6 = 4 monkeys; Row D

Page 36
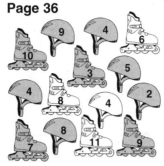

18 ÷ 9 = 2; 10 miles; 56 ÷ 8 = 7; 48 ÷ 8 = 6; 40 ÷ 8 = 5; 80 ÷ 8 = 10

Page 37
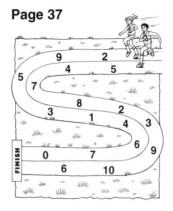

36 ÷ 4 = 9 miles

Page 38
A. 3 x 4 = 12, 4 x 3 = 12, 12 ÷ 3 = 4, 12 ÷ 4 = 3; B. 6 x 7 = 42, 7 x 6 = 42, 42 ÷ 6 = 7, 42 ÷ 7 = 6; C. 3 x 9 = 27, 9 x 3 = 27, 27 ÷ 3 = 9, 27 ÷ 9 = 3; D. 4 x 5 = 20, 5 x 4 = 20, 20 ÷ 4 = 5, 20 ÷ 5 = 4; E. 7 x 8 = 56, 8 x 7 = 56, 56 ÷ 7 = 8, 56 ÷ 8 = 7; F. 3 x 6 = 18, 6 x 3 = 18, 18 ÷ 3 = 6, 18 ÷ 6 = 3

Page 39
A. 1 R4, 4 R1; B. 6 R2, 9 R1, 7 R5, 6 R5; C. 7 R2, 7 R1, 7 R1, 6 R6; D. 5 R1, 6 R2, 3 R1, 5 R2; 36 ÷ 6 = 6 hurdles

Page 40
A. 8 R1, 8 R4, 6 R1, 9 R2; B. 7 R1, 6 R1, 5 R3, 5 R1; C. 4 R1, 5 R1, 8 R2, 7 R1; D. 3 R2, 3 R1, 9 R2, 5 R6; E. 4 R1, 6 R3, 2 R4, 6 R2; 56 ÷ 8 = 7 apples

Page 41
A. 14 ÷ 3 = 4 R2, 4 sweaters; B. 57¢ ÷ 9 = 6 R3, 6 candy canes; C. 71 ÷ 8 = 8 R7, 7 cookies; D. 17 ÷ 3 = 5 R2, 2 cars; E. 74 ÷ 8 = 9 R2, 2 cans; F. 85 ÷ 9 = 9 R4, 9 groups

Page 42
A. 22, 24, 31, 13; B. 32, 21, 34, 10; C. 33, 23, 40, 11

Page 43
A. 29, 17, 12, 18; B. 16, 12, 37, 23; C. 13, 19, 13, 15; 87 ÷ 3 = 29 marbles

Page 44
A. 63, 24, 36, 35; B. 9, 18, 48, 81; C. 32, 21, 16, 18; D. 64, 9, 30, 7; E. 160, 90, 300; F. 420, 300, 80; 128 people

Page 45
A. 96 ÷ 6 = 16 cookies; B. 98 ÷ 9 = 10 candy bars, 8¢ left over; C. 72 ÷ 4 = 18 bags of jelly beans; D. 84 ÷ 6 = 14 cookies; E. 63 ÷ 5 = 12 R3, 12 candy canes; F. 74 ÷ 7 = 10 R4, 10 pieces of licorice; 1 bag of jelly beans; candy cane and jelly beans

Page 46
A. 75 ÷ 3 = 25, 96 ÷ 4 = 24, Ellen; B. 18 x 9 = 162 times; C. 4 x 20 = 80, 80 ÷ 4 = 20 4-oz. servings; D. 4 x 12 = 48 eggs; E. 81 ÷ 9 = 9 bracelets; F. $.50 x 8 = $4